Math Mammoth
Place Value 1

By Maria Miller

Contents

Introduction 5

Games and Activities 6

Counting in Groups of 10 9

Naming and Writing Numbers 11

The "Teen" Numbers 15

Building Numbers 11-40.................................. 18

Building Numbers 41-100 20

A 100-Chart 22

Add and Subtract Whole Tens 24

Practicing with Numbers 26

Which Number Is Greater? 28

Even and Odd Numbers 31

Numbers Past 100 33

More Practice with Numbers 35

Skip-Counting Practice 37

Rounding to the Nearest Ten 40

Bar Graphs 42

Tally Marks 44

Regrouping: Make Another Ten
from the Ones 46

Review 50

Answers 53

Printable Number Chart 67

Printable Number Lines 69

More from Math Mammoth 71

Introduction

In this book, children learn numbers up to 120. They compare whole numbers within 100, and learn to think of them in terms of tens and ones.

When children count, they basically just learn numbers as some kind of continuum that goes on without end. With simple counting, your child might not catch on to the inherent structure of the number system. Our number system is based on the idea that if you have lots and lots of objects, the efficient way to count and denote them is with *groups* of tens, hundreds, and thousands — not individually.

The crucial point in understanding the concept of place value is therefore that a **certain position represents a group of a specific size**. The digit in each position tells us how many groups of that size there are. For example, in the number 2,381, an adult already knows that the 8 represents eight tens, and not just "8" and that the 3 represents three hundreds, and not just "3". The place of the digit tells us the size of the group, and the digit itself tells how many of that group.

The initial lessons of the book that introduce tens and ones use a **100-bead abacus** extensively. This 100-bead abacus or school abacus simply contains ten beads on ten rods, for a total of 100 beads. It is not the special abacus used by the Chinese or the Japanese. Each bead simply represents one. The 100-bead abacus lets children both "see" the numbers and use their touch while making them.

You will need to purchase this school abacus separately, such as on Amazon, or make your own.

You can browse Amazon's abacus collection at this link:
https://www.amazon.com/s?k=abacus+100+beads&tag=mathmammoth-20

Instead of a physical abacus, you can use this online virtual abacus:
https://apps.mathlearningcenter.org/number-rack/

Or, you can make one on your own. This is a fairly easy craft project and you can easily find instructions for it on the Internet (search for example for "DIY abacus").

Besides the abacus, we also use a visual model of blocks where ten of them "snap" together to form a stick. If you already have these so-called base-ten blocks, you can use them along with the visual exercises, if you prefer.

Moreover, we also use number lines and a 100-chart. Number lines help visualize how numbers continue indefinitely and also relate to the concept of measuring. The 100-chart helps the child to be familiar with the numbers below 100 and find patterns in the number system.

While most of the lessons in the book focus on place value, students also practice adding and subtracting multiples of ten and skip-counting. The two lessons at the end of the chapter about tally marks and graphs are real-life applications of two-digit numbers.

I wish you success in teaching math!

Maria Miller, the author

Games and Activities

Big and Bigger!

You need: A standard deck of playing cards from which you remove the face cards and tens, leaving only numbers from 1 through 9.

Game Play: In each round, each player is dealt *three* cards, and they make a two-digit number using *two* of the cards, however they want. The player with the largest two-digit number then wins all the cards used in that round. If there is a tie, those players draw two additional cards each, and resolve the tie with those.

After enough rounds have been played to use all of the cards, the player with the most cards wins.

Variations:
* The player with the smallest number wins each round.
* The goal is to make a number that is as close as possible to 50.
* Each player is dealt 4 cards, and they choose two of them to make their two-digit number.
* Each player is dealt 4 cards, they form two 2-digit numbers out of them, and add those. The player with the largest (and correct) sum wins the cards. This may be too challenging, but you can make it a bit easier by removing 9s, 8s, 7s, and 6s from the deck.

Abacus

The lessons in the book include several activities with the 100-bead abacus. Follow the instructions in the lessons.

Order Us!

You need: Two standard decks of playing cards from which you remove the face cards and tens, leaving only numbers from 1 through 9. Other number cards work as well. Paper and pencil to keep track of points.

Game Play: Place the deck of cards face down in a pile in the middle. At your turn, draw six cards. You will then form three 2-digit numbers using the cards any way you wish, and lastly order those from smallest to largest. If they are ordered correctly, you keep the cards and place them into your personal pile.

After enough rounds have been played to use all of the cards, the player with the most cards wins.

Variations:
* Each player draws 8 cards instead of 6, and makes four 2-digit numbers.

Games and Activities at Math Mammoth Practice Zone

Place Value
Choose Max Digits 2 to get 2-digit numbers.
https://www.mathmammoth.com/practice/place-value

Skip-count in a 100-chart
In this activity, the child fills in five numbers on a 100-chart in a specific skip-counting pattern. This helps the child to understand what skip-counting is, and to see different patterns that skip-counting forms on a 100-chart. The options allow you to change the starting and ending number of the grid. You can also choose the percentage of numbers that are already pre-filled in the chart.
https://www.mathmammoth.com/practice/skip-count-hundred-chart

Give the Cat a Mouse
In this game, the child is asked to find a specific number on a 100-chart. The options allow you to change the starting numbers and the step, allowing you to create, not only a chart from 1-100, but others such as a chart from 0 to 990 by tens, or a chart from 1 to 199 by twos, etc. You can also choose how many numbers the child will find, and the percentage of numbers that are already pre-filled in the chart.
https://www.mathmammoth.com/practice/hundred-chart

Ordering Numbers — Online Game
In this activity, children practice ordering numbers by dragging them into the given boxes in the correct order. Choose two-digit numbers.
https://www.mathmammoth.com/practice/order-numbers

Plot numbers on the number line
In this activity, the child drags a dot to position the given number on a number line, practicing place value and estimation. In the "normal" mode, you see plenty of tick marks on the number line. In the "Estimation" mode, there are only a few tick marks. Choose two-digit numbers.
https://www.mathmammoth.com/practice/number-line

Fruity Math
Add a multiple of ten and a single-digit number (such as 2 + 70). Click the fruit with the correct answer and try to get as many points as you can within two minutes.
https://www.mathmammoth.com/practice/fruity-math#op=addition&duration=120&mode=manual&config=1,9x1__1,9x10&max-sum=120

Further Resources on the Internet

We have compiled a list of external Internet resources that match the topics in this book. This list of links includes web pages that offer:

- **online practice** for concepts;
- online **games**, or occasionally, printable games;
- **animations** and interactive **illustrations** of math concepts;
- **articles** that teach a math concept.

We heartily recommend you take a look at the list. Many of our customers love using these resources to supplement the bookwork. You can use the resources as you see fit for extra practice, to illustrate a concept better, and even just for some fun. Enjoy!

https://l.mathmammoth.com/blue/placevalue1

Scan me

8

Counting in Groups of 10

1. Count in groups of TEN. Count ten dots, and circle them.
   Write how many groups of ten you get. Write how many ones are left over.

a.

groups of ten | ones

b.

groups of ten | ones

c.

groups of ten | ones

d.

groups of ten | ones

e.

groups of ten | ones

f.

groups of ten | ones

2. **Counting game 1.** (Optional - to give more practice for making groups of ten.) Put some beans or other objects onto a table. Ask the child to make groups of ten. Then ask the child to count the groups of ten and the individual items left over, using the form "four tens and six" or "seven tens and one", *etc*. Repeat as necessary with different amounts of objects, taking turns.

3. **Counting game 2.** You need: Counting items, such as sticks, beads, or beans. Containers, such as small bags or bowls. (If using sticks, then use rubber bands to group them, rather than containers.)

Before the game: Place 10 of the items in the middle and the rest in a pile on the side.

Play: At their turn, each player adds one more item to the middle pile on the table, and names the number that is formed. Whenever a whole ten is completed, those ten items are grouped together with a rubber band or by placing them in a small bag or bowl.

Use ONLY the number words from one to ten when counting in this game. Words like eleven, thirteen, fifty, *etc.* are not allowed. For example, eleven is said as "ten and one", twelve is "ten and two", twenty is "two tens", twenty-five is "two tens and five", *etc.*

Variation:
* Each player adds *two* (or more) items to the pile instead of just one.

4. Introduce the 100-bead abacus to the student.
   Make these numbers with the 100-bead abacus.

a. 6 tens, 5 ones	**e.** 2 tens, 1 one	**i.** 4 tens, 6 ones
b. 2 tens, 7 ones	**f.** 8 tens, 9 ones	**j.** 6 tens
c. 7 tens	**g.** 9 tens, 3 ones	**k.** 7 tens, 1 one
d. 1 ten, 5 ones	**h.** 1 ten, 1 one	**l.** 1 ten, 8 ones

5. Take turns telling each other what number to make on the abacus, such as "7 tens and 9" or "1 ten and 7". **Do not** proceed farther until the student has mastered this! This is crucial.

The names of the numbers with whole tens are:

ten	=	ten	four tens	=	forty	seven tens	=	seventy
two tens	=	twenty	five tens	=	fifty	eight tens	=	eighty
three tens	=	thirty	six tens	=	sixty	nine tens	=	ninety

ten tens = one hundred

---

6. Say the number names from ten to a hundred aloud a few times while also making the numbers on the 100-bead abacus. It almost sounds like a rhyme!

Naming and Writing Numbers

Use the names for the whole tens (twenty, thirty, forty, *etc.*) when naming numbers. Write the number of tens and the number of ones next to each other with numerals.

"6 tens and 7 ones" is sixty-seven, or 67.   "4 tens and 5 ones" is forty-five, or 45.

"8 tens and 1 one" is eighty-one, or 81.   "2 tens and 2 ones" is twenty-two, or 22.

Numbers between 10 and 20 are more difficult. We will look at them in the next lesson.

1. Make these numbers with the 100-bead abacus. Also name them and write them with numbers.

   a. 5 tens 6 ones   *fifty-six*   56

   b. 7 tens 2 ones

   c. 2 tens 1 one

   d. 3 tens 1 one

   e. 4 tens 8 ones

   f. 3 tens 5 ones

   g. 2 tens 3 ones

   h. 8 tens 7 ones

   i. 9 tens 4 ones

   j. 6 tens 6 ones

2. The **Counting game again** - this time with number names from 20 on.

You need: Counting items, such as sticks, beads or beans.
Containers, such as small bags or bowls. (If using sticks, then use rubber bands to group them, rather than containers.)

Preparation: Place 20 of the items in the middle and the rest in a pile to one side.

Play: During their turn, each player adds one more item to the middle pile on the table, and names the number that is formed. Whenever a whole ten is completed, those ten items are grouped together with a rubber band or by placing them in a small bag or bowl.

Variation:
* Each player adds *two* (or more) items to the pile instead of just one.

3. Take turns telling each other what number to make on the abacus, such as "fifty-six". Do not use the numbers between 10 and 20 yet.

4. Make these numbers with the abacus or with some counting items. Fill in the table.

a. twenty-four   __2__ tens __4__ ones = __24__

b. fifty-three   _____ tens _____ ones = _____

c. thirty-nine   _____ tens _____ ones = _____

d. sixty-two   _____ tens _____ ones = _____

e. eighty-eight   _____ tens _____ ones = _____

f. thirty-one   _____ tens _____ ones = _____

g. seventy-eight   _____ tens _____ ones = _____

h. ninety-nine   _____ tens _____ ones = _____

i. twenty-two   _____ tens _____ ones = _____

j. twenty-nine   _____ tens _____ ones = _____

tens	ones
3	5

In the number 35, the "3" means three tens. The "3" is in the tens place. The "5" means five ones. The "5" is in the ones place.

tens	ones
7	0

In the number 70, there are just seven whole tens, and no ones, so write "7" in the tens place and a zero in the ones place.

5. Circle each group of ten dots. Then count the groups of ten and the ones left over. Write and name the number. Notice how quick it is to count the dots this way!

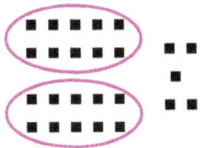

tens   ones

2	5

a. __twenty-five__

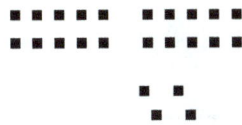

tens   ones

b. _____

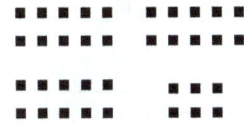

tens   ones

c. _____

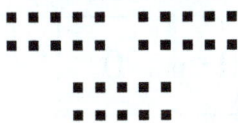

tens   ones

d. _____

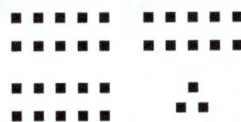

tens   ones

e. _____

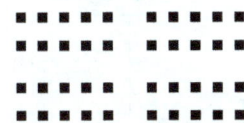

tens   ones

f. _____

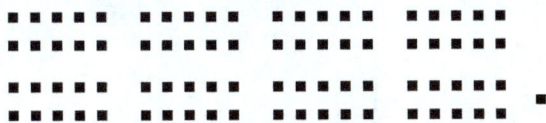

tens   ones

g. _____

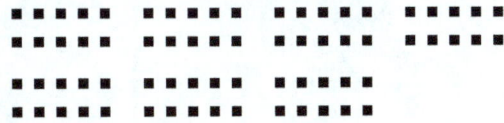

tens   ones

h. _____

6. Circle each group of ten dots. Then count the groups of ten and the ones left over. Write and name the number.

tens   ones

a. _____

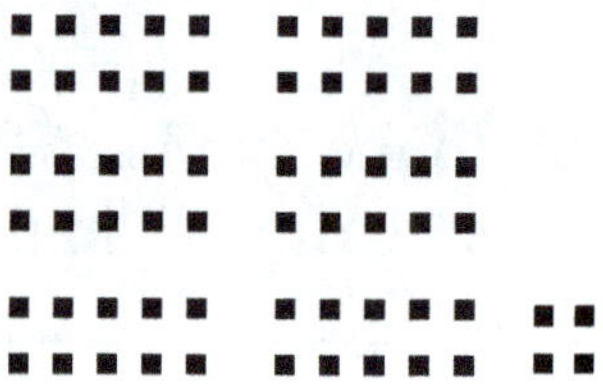

tens   ones

b. _____

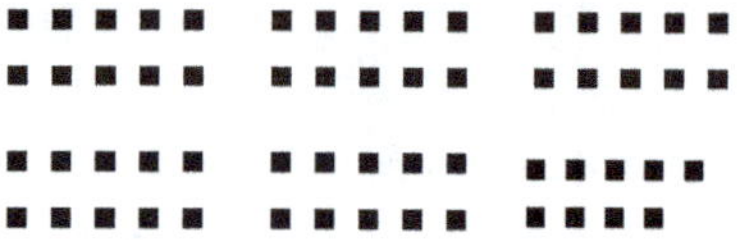

tens   ones

c. _____

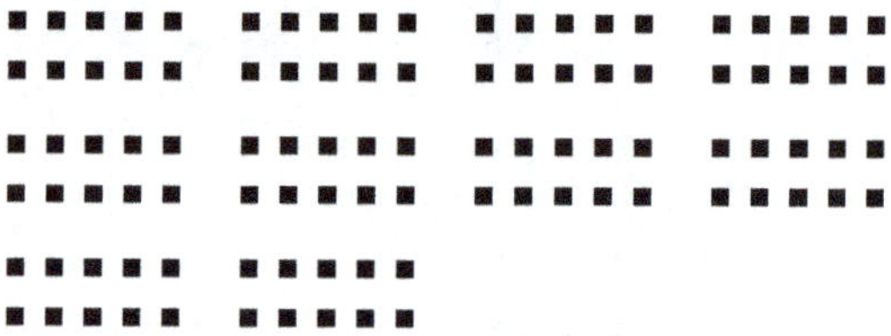

hundreds	tens	ones
1	0	0

d. *one hundred*

7. Now draw the dots yourself. Make groups of ten like before.

tens   ones

a. twenty-three

tens   ones

b. fifty

tens   ones

c. thirty-nine

The "Teen" Numbers

The names of the numbers between 10 and 20 are:

1 ten 1 = 11 = eleven   1 ten 6 = 16 = sixteen

1 ten 2 = 12 = twelve   1 ten 7 = 17 = seventeen

1 ten 3 = 13 = thirteen   1 ten 8 = 18 = eighteen

1 ten 4 = 14 = fourteen   1 ten 9 = 19 = nineteen

1 ten 5 = 15 = fifteen   2 tens   = 20 = twenty

The word "teen" actually comes from "ten". So, seventeen is actually "seven-ten", or one ten and seven.

1. Write the numbers on this number line under the tick marks. Say the numbers aloud.

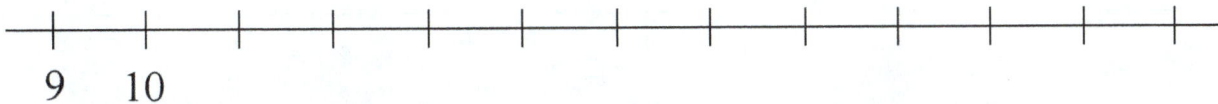

```
  ┼──┼──┼──┼──┼──┼──┼──┼──┼──┼──┼──┼──┼──┼
  9  10
```

2. Name and write the numbers.

   a. 1 ten 1 one _____ _____

   b. 1 ten 7 ones _____ _____

   c. 1 ten 2 ones _____ _____

   d. 1 ten 6 ones _____ _____

   e. 1 ten 3 ones _____ _____

   f. 1 ten 8 ones _____ _____

   g. 1 ten 4 ones _____ _____

   h. 1 ten 5 ones _____ _____

3. Fill in the missing items.

a. 10 + 8 = __18__ *eighteen*

b. 10 + 1 = _____ _____

c. 10 + _____ = 14 _____

d. 10 + _____ = 19 _____

e. _____ + 3 = _____ _____

f. 10 + 2 = _____ _____

g. _____ + _____ = 17 _____

h. _____ + _____ = 15 _____

4. This is a number chart from 1 to 100.

a. Find the whole tens on the chart (ten, twenty, thirty, and so on). Color their squares yellow.

b. Find the "teen" numbers on the chart (from thirteen to nineteen). Color their squares pink.

c. The "sixties" row is colored, ending in 70. Find the "forties" row and color it light green. Color the "nineties" row light brown.

d. Find the column that has all of the numbers that end in "5", such as 5, 15, 25, and so on. Color them light blue.

1	2	3	4	5	6	7	8	9	10
11	12	13	14	15	16	17	18	19	20
21	22	23	24	25	26	27	28	29	30
31	32	33	34	35	36	37	38	39	40
41	42	43	44	45	46	47	48	49	50
51	52	53	54	55	56	57	58	59	60
61	62	63	64	65	66	67	68	69	70
71	72	73	74	75	76	77	78	79	80
81	82	83	84	85	86	87	88	89	90
91	92	93	94	95	96	97	98	99	100

5. Choose one (or both) of the following games to practice the teen numbers some more.

Practice "Teens" board game

You need: A board game where you move a piece along a path. Number cards 1 to 9 plus one card with 10, from a standard deck of playing cards or UNO cards, etc.

Before the game: Place the card with 10 right-side-up in the middle. Place the deck next to it, face down.

Play: To start their turn, a player draws a card from the deck, placing it next to the 10-card. Then they have to *name* the number formed by the 10-card and the card they drew. For example, if 7 is drawn, the player has to say "seventeen". If the player names the number correctly, they can move their piece the number of steps that was on the card they drew (not including the ten). Otherwise, follow the rules of the board game you are using.

Teens out!

You need: A standard deck of playing cards. Each face card (jack, queen, king) counts as a 10.

Before the game: Deal each person 7 cards. Place the remaining cards (the deck) face down in the middle.

Play: To start their turn, the player takes one card from the deck. Then they discard from their hand any two cards that when added together are more than 10 but less than 20, *and* names the number when putting the cards down. For example, 10 and 5 make fifteen and can be discarded. Eight and five make thirteen, and can be discarded. Remember, any face card counts as a 10. A player may discard several pairs of cards in one turn. If the player fails to name the number correctly when discarding, they have to keep the cards in their hand. The game ends when someone is able to discard all of their cards from their hand.

Variation:
* Instead of discarding the cards, the player places a "teen" pair of cards face up in front of them as a "book". The game is played till all the cards are used from the deck. If a person does not have any cards left in their hand, then at their turn, they simply draw one card from the deck, and the turn passes to the next person. In the end, whoever has the most "books" wins.

Building Numbers 11-40

1. Fill in the table. Think of the " + " sign as "and": 10 + 3 means 10 *and* 3.

a. _____ eleven _____

__10__ + __1__

tens	ones
1	1

b. _____

_____ + _____

tens	ones

c. _____

_____ + _____

tens	ones

d. _____

_____ + _____

tens	ones

e. _____

_____ + _____

tens	ones

f. _____ Twenty _____

20 + __0__

tens	ones

g. _____ Twenty-one _____

20 + _____

tens	ones

h. _____

20 + _____

tens	ones

i. _____

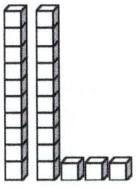

_____ + _____
tens   ones

j. _____

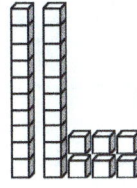

_____ + _____
tens   ones

k. _____

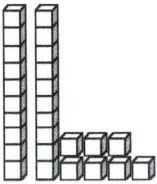

_____ + _____
tens   ones

l. _____

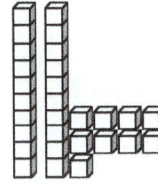

_____ + _____
tens   ones

m. _____

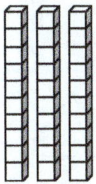

__30__ + __0__
tens   ones

n. _____

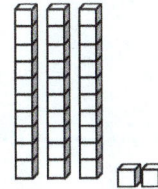

_____ + _____
tens   ones

o. _____

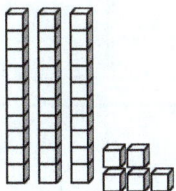

_____ + _____
tens   ones

p. _____

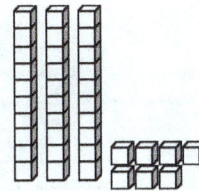

_____ + _____
tens   ones

q. _____

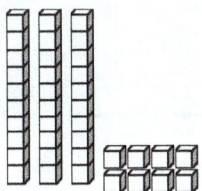

_____ + _____
tens   ones

r. _____

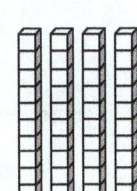

_____ + _____
tens   ones

Building Numbers 41-100

1. Fill in the table.

a. <u>**Forty-one**</u> _____ + _____ tens ones [ ][ ]	**b.** _____ _____ + _____ tens ones [ ][ ]
c. _____ _____ + _____ tens ones [ ][ ]	**d.** _____ _____ + _____ tens ones [ ][ ]
e. _____ _____ + _____ tens ones [ ][ ]	**f.** _____ _____ + _____ tens ones [ ][ ]
g. _____ _____ + _____ tens ones [ ][ ]	**h.** _____ <u>100</u> + <u>0</u> + <u>0</u> hundreds tens ones [1][0][0]

2. Match the number to its name.

46	forty-two	99	seventy-six
64	forty-six	81	eighty-one
55	fifty-five	90	ninety-one
70	fifty-seven	91	eighty-three
69	fifty-nine	79	ninety
59	sixty-four	76	seventy-nine
42	seventy	100	ninety-nine
57	sixty-nine	83	one hundred

3. Break the numbers into tens and ones.

a. 73 = __70__ + __3__   c. 45 = _____ + _____   e. 98 = _____ + _____

b. 91 = _____ + _____   d. 83 = _____ + _____   f. 64 = _____ + _____

4. Do the same thing the other way around! Add.

a.	b.	c.	d.
60 + 7 = _____	4 + 50 = _____	6 + 80 = _____	90 + 9 = _____
80 + 0 = _____	8 + 80 = _____	0 + 40 = _____	1 + 60 = _____

5. Fill in the missing numbers on the number lines.

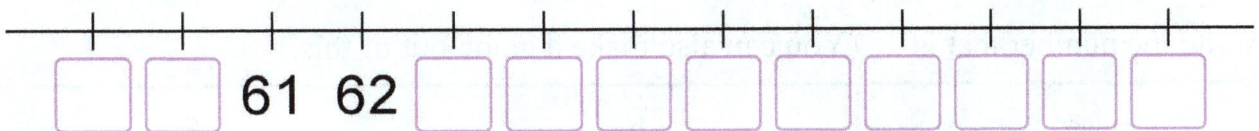

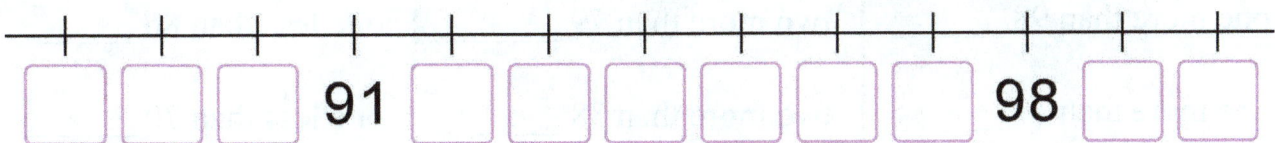

[ ] [ ] 61 62 [ ] [ ] [ ] [ ] [ ] [ ] [ ]

[ ] [ ] [ ] 91 [ ] [ ] [ ] [ ] [ ] 98 [ ] [ ]

A 100-Chart

1. Fill in the numbers missing from this 100-chart.

1		3		5				9	10
11	12							19	
21		23		26					
						38			
41			44						50
	53			56			59		
61			64		67				
	73				77		79		
81			84						
						97	98		100

2. Find these numbers on the chart: 26, 58, 34, 91, 70, and 45. You can color the squares.

3. Practice finding numbers on the chart with your friend or teacher. Take turns giving each other a number to find. Or, play this online game:

   Give the Cat a Mouse—Find the hidden mouse on a 100-chart.
   https://www.mathmammoth.com/practice/hundred-chart

4. Write the number that is...   (You can also make a game out of this.)

a.	b.	c.
one more than 98 _____	two more than 79 _____	one less than 86 _____
one more than 27 _____	two more than 38 _____	one less than 70 _____

5. Look at the row that starts with 31.
   How are those numbers alike?

6. Look at the column that starts with 5.
   How are those numbers alike?

7. For each highlighted blue number,
   find the number that is exactly
   ten more, and color it.
   Hint: It is exactly under it, in the next row.

8. For each underlined number,
   find the number that is exactly
   ten less, and underline it.
   Hint: It is exactly above it, in the previous row.

1	2	3	4	5	6	7	8	9	10
11	12	**13**	14	15	16	17	18	19	20
21	22	23	24	25	26	**27**	28	29	30
31	32	33	34	35	36	37	38	39	**40**
41	42	43	44	45	46	47	48	49	50
51	52	53	54	55	56	**57**	58	59	60
61	62	63	64	65	66	67	68	69	70
71	72	73	74	75	76	77	78	79	80
81	82	83	**84**	85	86	87	88	89	90
91	92	93	94	95	96	97	98	**99**	100

What number is 10 more than 49? In other words, what is 49 + 10?
Simply look at the tens number: **4**9, and add one to it. You get 59.

What number is 10 less than 72? In other words, what is 72 − 10?
 Look at the tens number: **7**2, and take away one from it. You get 62.

---

9. Write the number that is...

a.	b.	c.
ten more than 18 _____	ten more than 60 _____	ten less than 24 _____
ten more than 35 _____	ten more than 59 _____	ten less than 37 _____

10. Fill in the numbers missing from these number lines.

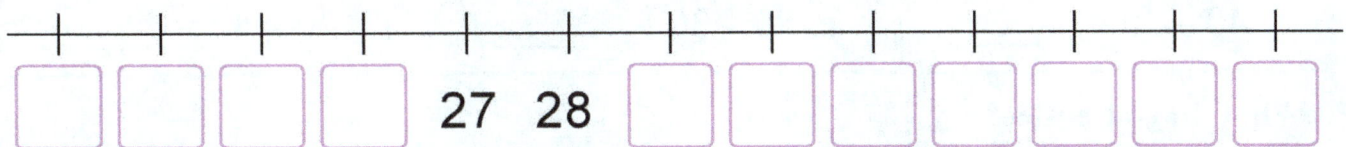

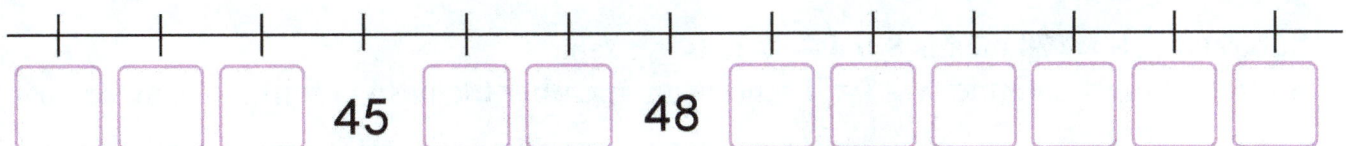

[number line with boxes: □ □ □ □ 27 28 □ □ □ □ □ □ □]

[number line with boxes: □ □ □ 45 □ □ 48 □ □ □ □ □ □]

Add and Subtract Whole Tens

1. Add and subtract with tens. You can also solve these with the abacus.

a. 30 + 20 = _____

b. 50 + 30 = _____

c. 40 + 40 = _____

d. 20 + 70 = _____

e. 70 + 30 = _____

f. 30 + 30 = _____

g. 30 − 10 = _____

h. 50 − 30 = _____

i. 70 − 40 = _____

j. 40 − 30 = _____

k. 80 − 30 = _____

l. 100 − 40 = _____

What do you notice?

Adding **60 + 30** means **6 tens + 3 tens** = 9 tens.
So I can simply add 6 + 3 = 9, and just remember the answer will be 9 tens, or 90.

Subtracting **80 − 70** means **8 tens − 7 tens** = 1 ten.
So I can simply subtract 8 − 7 = 1, and just remember the answer will be 1 ten, or 10.

2. Now, try these *without* using the abacus or other help.

a. 10 + 50 = _____ 60 + 20 = _____	**b.** 60 + 30 = _____ 20 + 50 = _____	**c.** 70 − 10 = _____ 90 − 40 = _____

3. Subtract from 100. Use the abacus if you need to.

a. 100 − 50 = _____ 100 − 20 = _____	**b.** 100 − 30 = _____ 100 − 70 = _____	**c.** 100 − 90 = _____ 100 − 40 = _____

4. Fill in the missing numbers.

a. 60 + ☐ = 90 20 + ☐ = 100	**b.** 80 − ☐ = 30 80 − ☐ = 40	**c.** ☐ − 10 = 10 ☐ − 20 = 20

5. Add the same number several times.

a. 10 + 10 + 10 + 10 + 10 = _____   **b.** 20 + 20 + 20 + 20 + 20 = _____

6. Add and subtract whole tens.

a. 20   +30 →☐ +10 →☐ −10 →☐ −50 →☐

b. 70   −30 →☐ +10 →☐ +20 →☐ −40 →☐

Puzzle Corner   *Quick additions!* In these sums, first add any numbers that make ten. (We can do that because we can add in any order.)

10 + 2 + 20 + 4 + 5 + 1 + 8 + 9 + 9 + 20 + 5 + 6 = _____

8 + 6 + 7 + 9 + 5 + 30 + 4 + 3 + 5 + 2 + 1 + 20 + 1 = _____

Practicing with Numbers

1. Fill in.

a. _____

_____ + _____
tens   ones

b. _____

_____ + _____
tens   ones

c. _____

_____ + _____
tens   ones

d. _____

_____ + _____
tens   ones

2. Add ten by drawing another column of ten. Subtract ten by crossing out a column.

a. $43 + 10 =$ _____

b. $28 + 10 =$ _____

c. $56 - 10 =$ _____

3. *Explain* to your teacher how you add a ten to a number, or subtract a ten.

a. $61 + 10 =$ _____

$61 - 10 =$ _____

b. $37 + 10 =$ _____

$37 - 10 =$ _____

c. $89 + 10 =$ _____

$89 - 10 =$ _____

4. Match the number to its name.

16	twelve	90	eleven
60	fourteen	29	nineteen
23	sixteen	11	ninety-three
32	thirty-two	19	ninety-one
76	sixty	91	twenty-nine
14	twenty-three	39	thirty-nine
12	seventy-six	93	ninety

5. Write each number as (tens) + (ones).

a. 91 = _____ + _____   b. 79 = _____ + _____   c. 58 = _____ + _____

6. Add.

a. 20 + 6 = _____   b. 60 + 7 = _____   c. 6 + 50 = _____

80 + 2 = _____   10 + 1 = _____   5 + 30 = _____

7. Fill in.

[ ] [ ] [ ] 35 [ ] [ ] [ ] [ ] [ ] 42 [ ] [ ]

[ ] [ ] [ ] [ ] [ ] 32 [ ] [ ] [ ] [ ] [ ] [ ]

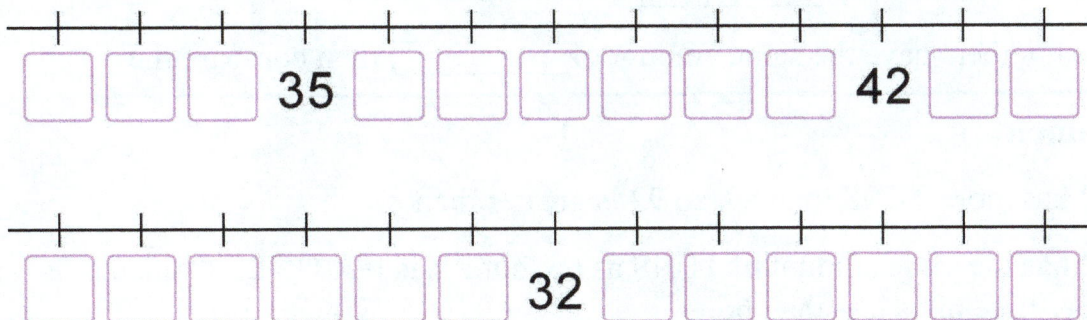

Mystery Number
38 25 31 1 99 47 101 99

I'm between 84 and 92. I don't have any ones. Who am I?

Which Number is Greater?

1. The pictures have ten-sticks and one-dots. Fill in the missing parts.
   Then circle the number that is **more**.

a. ||| ...... OR |||||| ...

   __3__ tens __6__ ones   ____ tens ____ ones

   __36__   _____

b. ||||| .... OR |||| .....

   __54____   _____

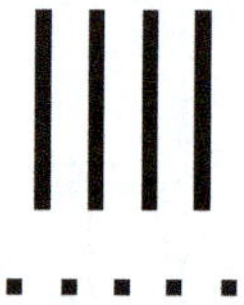

c. |||||| .... OR ||||||| ......

   ____ tens ____ ones   ____ tens ____ ones

   _____   _____

d. |||| ...... OR |||||| ....

   _____   _____

Study the above pictures. Do we first check how many TENS the numbers have or how many ONES the numbers have?

- Check first how many _____ the numbers have.

- If the numbers have the same amount of _____, then compare the _____.

For example:

- 92 has more TENS than 89, so 92 is greater than 89.

- 62 has the same amount of TENS as 66, but it has less ONES than 66. Therefore 62 is less than 66.

The symbol > means "greater than", and < means "less than". Think of it as a hungry alligator's mouth! The open end (the open "mouth") of the symbol ALWAYS points to the **bigger** number—the alligator wants to eat as much as it can!

For example: 3 < 5   14 > 3   60 > 50   48 < 99   7 < 17

2. Write < or > between the numbers. You can draw ten-sticks and one-dots to help.

|||| |||||
. . . .

a. 45 ☐ 54

b. 34 ☐ 24

c. 50 ☐ 54

d. 15 ☐ 56

e. 29 ☐ 64

f. 81 ☐ 90

g. 77 ☐ 47

h. 34 ☐ 94

i. 80 ☐ 68

3. Write < or > between the numbers to compare them. Number line can help.

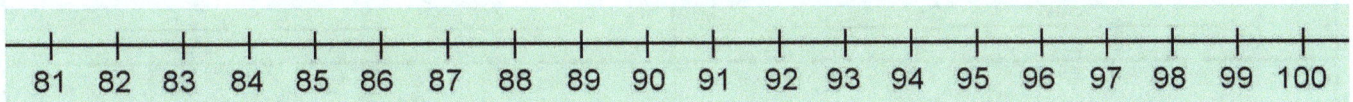

81 82 83 84 85 86 87 88 89 90 91 92 93 94 95 96 97 98 99 100

a. 81 < 88   b. 95 ☐ 59   c. 99 ☐ 96   d. 85 ☐ 91

e. 90 ☐ 88   f. 49 ☐ 94   g. 90 ☐ 99   h. 100 ☐ 87

i. 50 ☐ 55   j. 22 ☐ 12   k. 98 ☐ 89   l. 24 ☐ 42

4. Write $<$, $>$ or $=$.

a. $80 + 2$ $\boxed{<}$ $80 + 7$ b. $10 + 7$ $\boxed{\phantom{<}}$ $70 + 5$

c. $60 + 5$ $\boxed{\phantom{<}}$ $40 + 5$ d. $50 + 7$ $\boxed{\phantom{<}}$ $70 + 3$

e. $6 + 60$ $\boxed{\phantom{<}}$ $6 + 80$ f. $5 + 40$ $\boxed{\phantom{<}}$ $40 + 5$

g. $40 + 5$ $\boxed{\phantom{<}}$ $40 + 4$ h. $70 + 5$ $\boxed{\phantom{<}}$ $5 + 70$

5. Put the numbers in order from the smallest to the largest.

a. 68, 67, 46	b. 33, 53, 37
_____ < _____ < _____	_____ < _____ < _____
c. 48, 46, 18	d. 87, 78, 80
_____ < _____ < _____	_____ < _____ < _____
e. 48, 84, 46, 50	f. 98, 87, 67, 76
_____ < _____ < _____ < _____	_____ < _____ < _____ < _____

Mystery Number
38 2 1 99
47 101

Who am I?

I have two more tens than 40, and the same amount of ones as 17.

30

Even and Odd Numbers

Can John and Jane share 4 balls evenly (so that both get as many balls)?

● ● ● ●

Yes! Draw the balls for John and Jane.

John

Jane

Can John and Jane share 6 cars evenly? Try!

🚗 🚗 🚗 🚗 🚗 🚗

Can they share 5 carrots evenly? Try it!

🥕 🥕 🥕 🥕 🥕

Can they share 9 safety-pins evenly?

John

Jane

Four is an EVEN number because two people can share four things evenly.

Five is an ODD number because two people cannot share five things evenly.

1. Can two people share these things evenly? If yes, circle EVEN. If not, circle ODD.

10 marbles	7 marbles	3 stars
a. EVEN ODD	b. EVEN ODD	c. EVEN ODD
11 marbles	6 stars	4 marbles
d. EVEN ODD	e. EVEN ODD	f. EVEN ODD
9 stars	8 marbles	5 marbles
g. EVEN ODD	h. EVEN ODD	i. EVEN ODD

2. The chart shows how many cookies there are. Use rocks, beans, or other small items to make these amounts. Try to share them evenly with a friend. If you can share evenly, write "E" or "even" in the last column. If not, write "O" or "odd".

Cookies	Share evenly?	Even or odd?
11	NO	O
14		
15		

Cookies	Share evenly?	Even or odd?
12		
17		
16		

3. Color yellow all the EVEN numbers in the chart. Notice what pattern it makes! You can get help from your work in #1 and #2.

1	2	3	4	5	6	7	8	9	10
11	12	13	14	15	16	17	18	19	20

Now, color all the EVEN numbers in the rest of the 100-chart in the same pattern.

21	22	23	24	25	26	27	28	29	30
31	32	33	34	35	36	37	38	39	40
41	42	43	44	45	46	47	48	49	50
51	52	53	54	55	56	57	58	59	60
61	62	63	64	65	66	67	68	69	70
71	72	73	74	75	76	77	78	79	80
81	82	83	84	85	86	87	88	89	90
91	92	93	94	95	96	97	98	99	100

The numbers you didn't color are ODD numbers.

4. Look at the chart. Fill in.

Even numbers always end in (their last digit is) __2__, _____, _____, _____, or _____.

Odd numbers always end in (their last digit is) __1__, _____, _____, _____, or _____.

32

Numbers Past 100

Counting past 100 is very easy. You just start over, but instead of 1, 2, 3, 4, and so on, you count, "one hundred one", "one hundred two", "one hundred three", "one hundred four", and so on.

Count aloud with your teacher from one hundred to one hundred twenty.
Look at how these numbers are written:

100, 101, 102, 103, 104, 105, 106, 107, 108, 109, 110,
111, 112, 113, 114, 115, 116, 117, 118, 119, 120.

1. Fill in the missing numbers. This chart starts at one hundred one (101) and ends at 130.

101	102	103	104						110
111	112	113					118	119	120
121	122	123	124	125	126	127	128	129	130

2. Fill in the missing numbers. This chart starts at 81 and ends at 110.

81	82			85	86		88		
91			94			97			
		103			106			109	110

3. Write the number that is... (You can use the above charts to help.)

a.	b.	c.
one more than 109 _____	two more than 116 _____	ten more than 100 _____
one more than 113 _____	two more than 99 _____	ten more than 102 _____

d.	e.	f.
one less than 108 _____	two less than 110 _____	ten less than 120 _____
one less than 120 _____	two less than 117 _____	ten less than 107 _____

hundreds tens ones

hundreds	tens	ones
1	0	4

One hundred four blocks

hundreds	tens	ones
1	1	2

One hundred twelve blocks

4. Fill in the table.

a. _one hundred three_

hundreds	tens	ones
1	0	

b. _____

hundreds	tens	ones
1	0	

c. _____

hundreds	tens	ones

d. _____

hundreds	tens	ones

e. _____

hundreds	tens	ones

f. _____

hundreds	tens	ones

5. Fill in the missing numbers. You can also do this orally with your teacher.

[ ] [ ] [ ] [ ] 101 [ ] [ ] [ ] [ ]

[ ] [ ] [ ] [ ] [ ] [ ] [ ] 113 [ ]

34

More Practice with Numbers

1. Fill in the missing numbers on the chart.

31	32		34					39	
		43	44	45	46			49	
51	52	53				57			60
			64	65	66		68		
				76	77	78	79		
		83							
	92	93	94	95	96				
101				105					110
					116	117			120

2. Use the chart above to help, and write the number that is...

a.	b.	c.
ten more than 67 _____	ten more than 78 _____	ten less than 55 _____
ten more than 87 _____	ten more than 101 _____	ten less than 70 _____
ten more than 45 _____	ten more than 99 _____	ten less than 118 _____

3. Draw another column of ten in each picture. That is adding 10! Then write the addition sentence. You can also use an abacus.

a. 32 + 10 = _____	b. 44 + 10 = _____	c. 28 + 10 = _____

4. Add 10. Use the 100-bead abacus to help.

a. 51 + 10 = _____	b. 16 + 10 = _____	c. 67 + 10 = _____
d. 74 + 10 = _____	e. 65 + 10 = _____	f. 39 + 10 = _____

5. Add the same number several times. Write the addition sentence.

a.

_____ + _____ + _____ = _____

b.

_____ + _____ + _____ = _____

c.

_____ + _____ + _____ + _____ + _____ = _____

6. Count by ones to fill in the missing numbers.

a. 88, 89, _____, _____

b. _____, 16, _____, 18

c. _____, 62, 63, _____

d. _____, 108, _____, 110

36

Skip-Counting Practice

1. Count by twos. You can also do this orally with your teacher.

   a. 0, 2, 4, _____, _____, _____, _____, _____, _____, _____, _____

   b. 50, 52, _____, _____, _____, _____, _____, _____, _____, _____

   c. 1, 3, 5, _____, _____, _____, _____, _____, _____, _____, _____

   d. _____, _____, _____, _____, 35, 37, _____, _____, _____, _____

2. Fill in the chart.

   Then color every fourth number on the number chart starting at 4. It should create a pretty pattern.

1	2	3	4	5	6	7	8		
									20
	22	23	24						30
31	32	33	34	35	36	37	38		
41							48	49	50
51	52	53				57			
61			64	65			68		
				75	76	77			
	82	83					87	88	89
91	92	93			96	97	98	99	100

3. Count by fours. You can also do this orally with your teacher.

   a. 0, 4, _____, _____, _____, _____, _____, _____, _____, _____

   b. 52, 56, _____, _____, _____, _____, _____, _____, _____, _____

   c. 1, 5, _____, _____, _____, _____, _____, _____, _____, _____

4. Color every third number on the number chart starting at 3. It should create a pretty pattern.

1	2	3	4	5	6	7	8	9	10
11	12	13	14	15	16	17	18	19	20
21	22	23	24	25	26	27	28	29	30
31	32	33	34	35	36	37	38	39	40
41	42	43	44	45	46	47	48	49	50
51	52	53	54	55	56	57	58	59	60
61	62	63	64	65	66	67	68	69	70
71	72	73	74	75	76	77	78	79	80
81	82	83	84	85	86	87	88	89	90
91	92	93	94	95	96	97	98	99	100

5. Count by tens. Use the chart to help. Notice the patterns!

a. 0, 10, _____, _____, _____

b. 21, 31, _____, _____, _____

c. _____, 12, _____, 32, _____

d. _____, _____, 65, 75, _____

e. _____, 26, _____, 46, _____

f. _____, _____, _____, 89, 99

6. Add 10 to these numbers. Notice: you only change how many tens there are.

a. 61 + 10 = _____

   15 + 10 = _____

b. 34 + 10 = _____

   82 + 10 = _____

c. 69 + 10 = _____

   55 + 10 = _____

7. Count by fives. Use the chart to help. You can also do this orally with your teacher.

a. 0, 5, _____, _____, _____, _____, _____, _____, _____, _____

b. 1, 6, _____, _____, _____, _____, _____, _____, _____, _____

Mystery Number
38 25 31 99
47 101

I have as many tens as 15 and as many ones as 99. Who am I?

Try your own coloring patterns on this page!

1	2	3	4	5	6	7	8	9	10
11	12	13	14	15	16	17	18	19	20
21	22	23	24	25	26	27	28	29	30
31	32	33	34	35	36	37	38	39	40
41	42	43	44	45	46	47	48	49	50
51	52	53	54	55	56	57	58	59	60
61	62	63	64	65	66	67	68	69	70
71	72	73	74	75	76	77	78	79	80
81	82	83	84	85	86	87	88	89	90
91	92	93	94	95	96	97	98	99	100

1	2	3	4	5	6	7	8	9	10
11	12	13	14	15	16	17	18	19	20
21	22	23	24	25	26	27	28	29	30
31	32	33	34	35	36	37	38	39	40
41	42	43	44	45	46	47	48	49	50
51	52	53	54	55	56	57	58	59	60
61	62	63	64	65	66	67	68	69	70
71	72	73	74	75	76	77	78	79	80
81	82	83	84	85	86	87	88	89	90
91	92	93	94	95	96	97	98	99	100

Rounding to the Nearest Ten

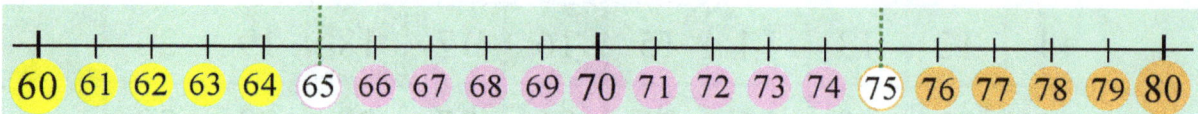

Find the whole tens on the number line. Which whole ten is 62 closest to? And 66? Which whole ten is 68 closest to? How about 77?

Number 65 is special: it is as close to 60 as it is to 70. What other number is like that?

Rounding a number means finding another, easy number that the number is close to. *Rounding a number to the nearest ten* means finding which whole ten the number is closest to.

We use the symbol ≈ when rounding. Read it as "is approximately" or "is about". So, 61 ≈ 60 can be read as "61 is approximately 60" or "61 is about 60".

1. Mark the numbers on the number line. Find which whole ten they are closest to.

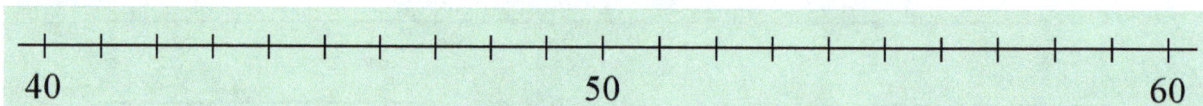

a. 52 is closest to _____   **b.** 57 is closest to _____   **c.** 43 is closest to _____

d. 48 is closest to _____   **e.** 59 is closest to _____   **f.** 46 is closest to _____

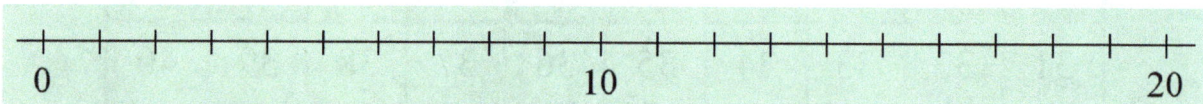

g. 7 is closest to _____   **h.** 3 is closest to _____   **i.** 14 is closest to _____

j. 19 is closest to _____   **k.** 12 is closest to _____   **l.** 4 is closest to _____

2. Round these numbers to the nearest whole ten. We will use the symbol " ≈ " this time.

a. 32 ≈ _____   **b.** 47 ≈ _____   **c.** 59 ≈ _____   **d.** 88 ≈ _____

e. 11 ≈ _____   **f.** 26 ≈ _____   **g.** 74 ≈ _____   **h.** 93 ≈ _____

The middle "5"

60	61	62	63	64	65	66	67	68	69	70	71	72	73	74	75	76	77	78	79	80

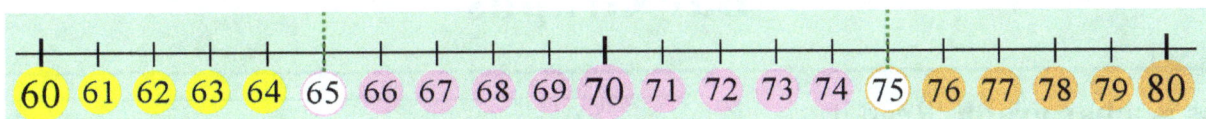

The "middle 5" is actually as far from the previous ten as it is from the next ten, but mathematicians have decided to *round it up*. This means we round any number ending in 5 to the "upper" or next greater whole ten. So, $65 \approx 70$ and $75 \approx 80$, and $5 \approx 10$.

And, *rounding down* would of course mean that you round a number towards the "lower" or previous, lesser ten-number. For example, 71 is rounded down to 70.

3. Round these numbers to the nearest whole ten.

    a. $35 \approx$ _____   **b.** $65 \approx$ _____   **c.** $95 \approx$ _____   **d.** $82 \approx$ _____

    e. $5 \approx$ _____   **f.** $66 \approx$ _____   **g.** $75 \approx$ _____   **h.** $38 \approx$ _____

So what good does it do to round? Any time that you do not *need* to know the exact answer, you can use rounded numbers to do calculations. We call this *estimation*.

Example. A cell phone costs $78 and another costs $51. So, the one costs *about* $80 and the other costs *about* $50. The one is *about* $30 more expensive than the other.

4. Find *about* how much the two things cost together. Use rounded numbers!

a.	b.	c.
a skirt, $28, and pants, $33	a bicycle, $56, and light, $12	a puzzle, $17, and book, $9
a skirt about $_____	bicycle about $_____	puzzle about $_____
pants about $_____	light about $_____	book about $_____
together about $ _____	together about $ _____	together about $ _____

5. A farmer has 49 sacks of apples and his neighbor has 18 sacks of apples. About how many sacks do they have all totaled?

6. About how much would a $13 DVD and two $18 DVDs cost together?

Bar Graphs

This is a **bar graph**. Read it this way: look at the TOP of each column (bar), and look towards the left. How high does the top of the bar reach? Read the number.

Look at the first bar, for short pencils. Where does the top of that bar reach?

It reaches to 7. So, Henry has 7 short pencils.

Henry's Pencils

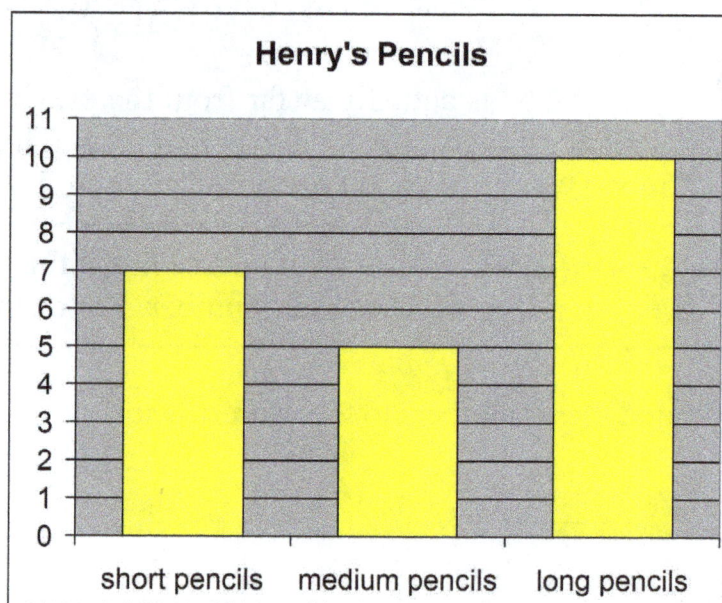

1. **a.** How many medium pencils does Henry have?

   b. How many long pencils does Henry have?

   c. How many short and medium pencils does Henry have in total?

   d. How many more long pencils does he have than short ones?

2. Here, the bar for first grade students reaches two little lines past 20. That's 22 students.

   a. How many students are in 2nd grade?

   b. How many students are in 3rd grade?

   c. How many students are in 4th grade?

Students

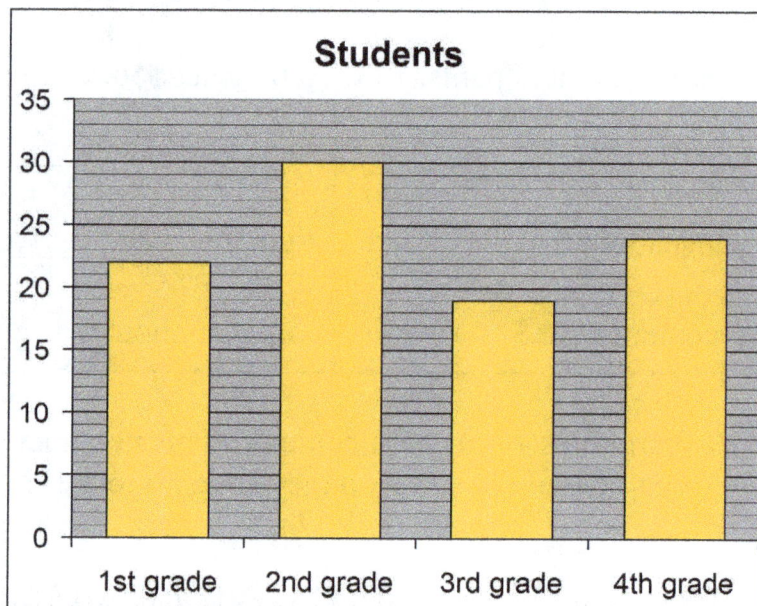

Books read during vacation

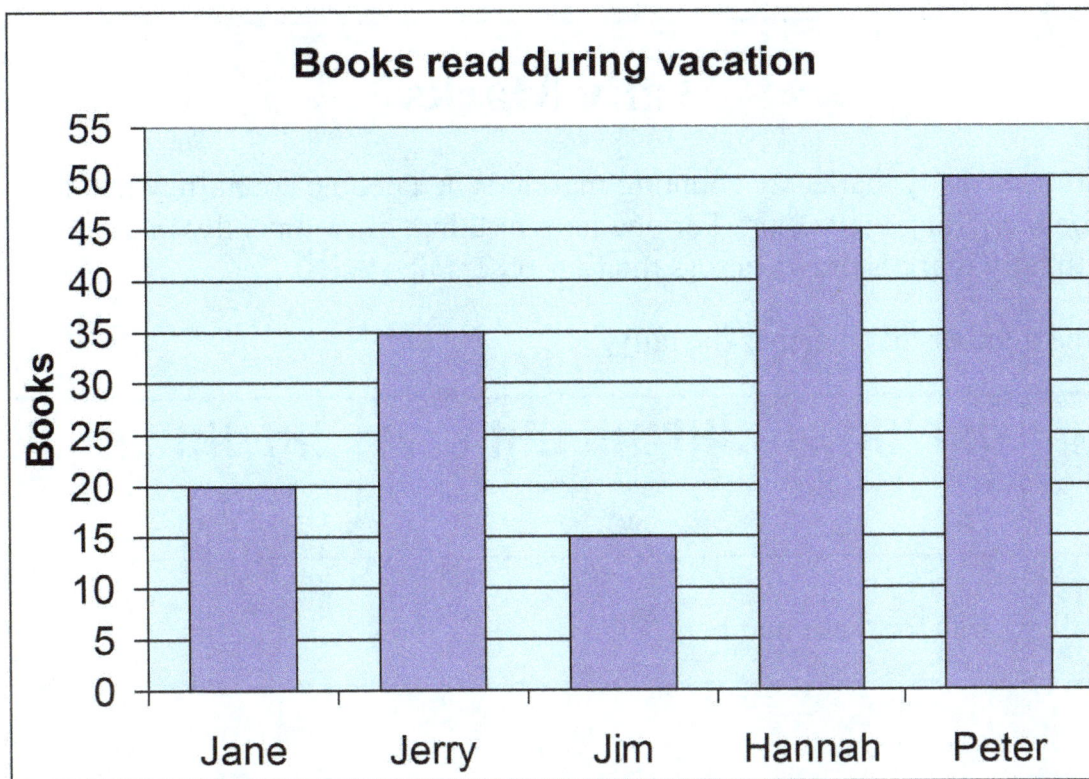

3. **a.** How many books did each child read?

Jane _____ Jerry _____ Jim _____ Hannah _____ Peter _____

b. _____ read the fewest books.

_____ read the most books.

c. The two children who read the most books were _____

and _____.

The two children who read the fewest books were _____

and _____.

d. How many books did Jane and Peter read in total? _____ books

(Challenge) How many total books did Jim and Hannah read? _____ books

Tally Marks

1. **Tally marks.** Tally marks are counting marks. When people count they make one tally mark for each thing they count. For one item or thing, draw one tally mark as " | ". The fifth tally mark is drawn across the four others like " ⅢⅠ ".

   Write the number that matches the tally.

ⅢⅠ	ⅢⅠ ⅢⅠ			ⅢⅠ ⅢⅠ ⅢⅠ					ⅢⅠ ⅢⅠ ⅢⅠ ⅢⅠ			
a. _____	b. _____	c. _____	d. _____

2. Draw tally marks for these numbers.

a. 7	b. 14
c. 16	d. 32
e. 41	f. 28

3. Count the fish. Use tally marks to keep track. Mark each fish you count and make a tally mark for it. That way you will not count the same fish twice. Then write the number under "Count".

	Tally Marks	Count
Red		
Blue		
Yellow		

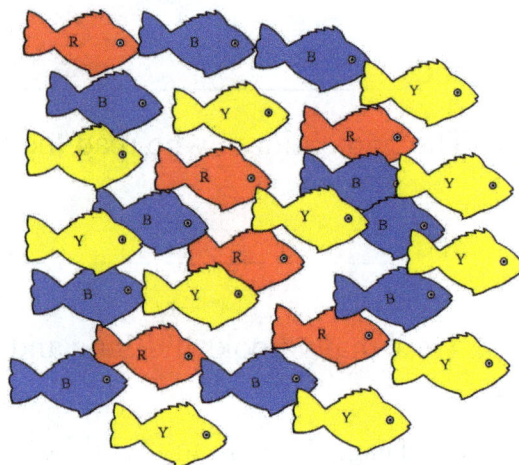

44

4. Count the pencils in each group. Use tally marks to keep track. Mark each pencil as you count it, and make a tally mark in the box. That way you will not count the same pencil twice.

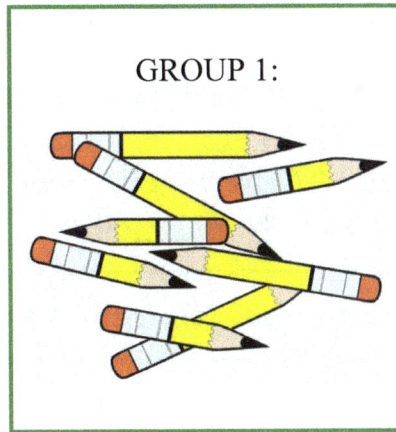

GROUP 1:

GROUP 2:

	Tally Marks	Count
Group 1		
Group 2		

5. Do the tally marks show the same counts that the bar graph does? If not, correct the tally.

	Tally Marks
Blue	ⅢⅢ ⅢⅢ ⅢⅢ ⅢⅢ ⅢⅢ II
Black	ⅢⅢ ⅢⅢ
White	ⅢⅢ ⅢⅢ ⅢⅢ ⅢⅢ II
Red	ⅢⅢ ⅢⅢ ⅢⅢ

Cars That Kyle Saw

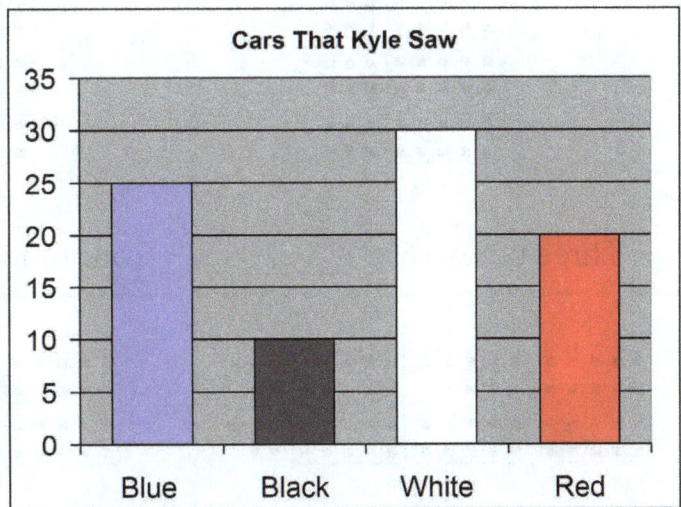

6. (Optional) Tally marks are most useful for counting things that are happening rather slowly, for example, birds that fly into the yard. For this project, count something using tally marks. For example, you could go outside and count how many red and how many gray cars you see pass by your house in 20 minutes.

	Tally Marks	Count
Group 1		
Group 2		

Regrouping: Make Another Ten from the Ones

The material in this lesson may be omitted. I have included it here for advanced students, and also to prepare the students for the idea of regrouping (carrying) that is found in second grade.

Make another ten from the ones

here is 14

and 14   =

and 14

In total we have
4 tens and 2 = 42.

14 + 14 + 14   =   42

---

1. Make a new group of ten from the ones. Count the tens. Count the ones.

a. Three times 16 = _____	**b.** Four times 13 = _____	**c.** Three times 23 = _____
d. Two times 35 = _____	**e.** Two times 28 = _____	**f.** Two times 17 = _____
g. Three times 36 = _____	**h.** Three times 25 = _____	**i.** Three times 15 = _____

2. Do the same as before, but now draw the groups of ten and ones yourself.

a. Two times 19 = _____

b. Two times 34 = _____

c. Three times 26 = _____

d. Four times 17 = _____

e. Three times 29 = _____

f. Two times 38 = _____

Make a new ten from the ones

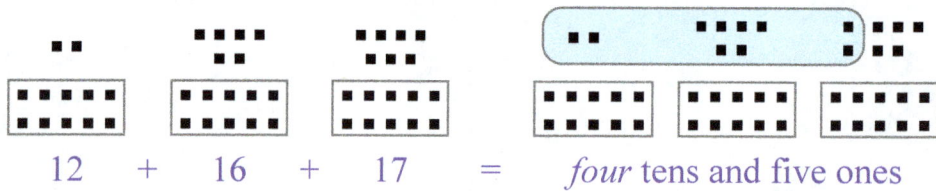

$12 + 16 + 17 =$ *four* tens and five ones

3. Add. Make a new ten from the ones.

a.

_____ + _____ + _____ = _____

b.

_____ + _____ + _____ = _____

c.

_____ + _____ = _____

d.

_____ + _____ = _____

e.

_____ + _____ + _____ = _____

f.

_____ + _____ + _____ = _____

48

4. Now make the pictures yourself! You can also use base ten blocks, an abacus, or other manipulatives.

a. 17 + 26 = _____

b. 23 + 28 = _____

c. 22 + 18 = _____

d. 19 + 12 + 16 = _____

e. 37 + 15 = _____

f. 20 + 13 + 19 = _____

Review

1. Name the numbers using numbers and words.

   a. 1 ten 5 ones   *15* _____

   b. 6 tens 7 ones   _____ _____

   c. 4 tens 0 ones   _____ _____

   d. 10 tens 0 ones   _____ _____

   e. 5 tens 1 one   _____ _____

2. Fill in the numbers missing from the number lines.

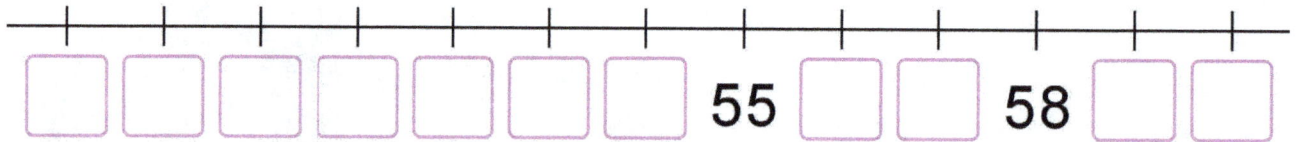

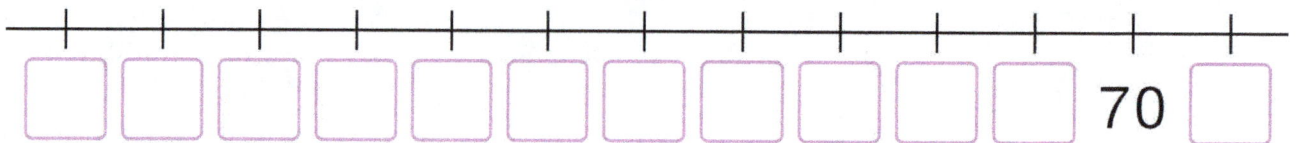

   ☐ ☐ ☐ ☐ ☐ ☐ ☐ **55** ☐ ☐ **58** ☐ ☐

   ☐ ☐ ☐ ☐ ☐ ☐ ☐ ☐ ☐ ☐ **70** ☐

3. Circle the number that is *more*.

a.	b.	c.	d.	e.
78 87	22 25	56 57	68 80	101 11

4. Count. You can also say this aloud with your teacher.

   97, 98, _____, _____, _____, _____, _____,

   _____, _____, _____, _____, _____, _____.

5. Break the numbers into their tens and ones.

a. 45 = 40 + 5	**b.** 25 = _____ + ____	**c.** 78 = _____ + ____
68 = _____ + ____	54 = _____ + ____	91 = _____ + ____

6. Build the numbers.

a. 50 + 7 = _____	**b.** 8 + 10 = _____	**c.** 90 + 6 = _____
20 + 0 = _____	9 + 70 = _____	9 + 60 = _____

7. Put the numbers in order.

a. 57, 17, 75	**b.** 18, 48, 44, 41
_____ < _____ < _____	_____ < _____ < _____ < _____

8. Compare the expressions and write < , > or =.

a. 56 ☐ 5 + 60   **b.** 20 + 8 ☐ 33   **c.** 60 + 5 ☐ 50 + 6

d. 34 ☐ 30 + 6   **e.** 4 + 90 ☐ 49   **f.** 80 + 2 ☐ 70 + 9

9. Skip-count. (You can also say this aloud with your teacher.)

a. 13, 15, 17, _____, _____, _____, _____, _____, _____

b. _____, _____, _____, _____, _____, _____, 78, 88, 98

c. _____, _____, _____, _____, _____, 55, 60, 65, _____

Mystery Number
38 26 31 99
47 101 9

I have five fewer ones than 39, and one more ten than 47.

51

Answer Key

Counting in Groups of 10, pp. 9-10

Page 9

1. a.

groups of ten	ones
2	4

b.

groups of ten	ones
4	2

c.

groups of ten	ones
5	6

d.

groups of ten	ones
5	2

e.

groups of ten	ones
6	3

f.

groups of ten	ones
6	9

Page 10

4. The teacher needs to check the student's work. For example, to make 6 tens and 5 ones in (a), move six full rows of beads and five beads from the seventh row on the abacus.

Naming and Writing Numbers, pp. 11-14

Page 11

1. a. fifty-six 56
   b. seventy-two 72
   c. twenty-one 21
   d. thirty-one 31
   e. forty-eight 48

   f. thirty-five 35
   g. twenty-three 23
   h. eighty-seven 87
   i. ninety-four 94
   j. sixty-six 66

Page 12

4. a. twenty-four 2 tens 4 ones = 24
   b. fifty-three 5 tens 3 ones = 53
   c. thirty-nine 3 tens 9 ones = 39
   d. sixty-two 6 tens 2 ones = 62
   e. eighty-eight 8 tens 8 ones = 88

   f. thirty-one 3 tens 1 one = 31
   g. seventy-eight 7 tens 8 ones = 78
   h. ninety-nine 9 tens 9 ones = 99
   i. twenty-two 2 tens 2 ones = 22
   j. twenty-nine 2 tens 9 ones = 29

Page 13

5. a. 25; twenty-five
   e. 33; thirty-three
   b. 24; twenty-four
   f. 40; forty
   c. 36; thirty-six
   g. 81, eighty-one
   d. 30; thirty
   h. 70; seventy

Page 14

6. a. 97; ninety-seven b. 64; sixty-four c. 59; fifty-nine d. 100; one hundred

7.

tens	ones
2	3

a. twenty-three

tens	ones
5	0

b. fifty

tens	ones
3	9

c. thirty-nine

The "Teen" Numbers, pp. 15-17

Page 15

1.

| 9 | 10 | 11 | 12 | 13 | 14 | 15 | 16 | 17 | 18 | 19 | 20 | 21 |

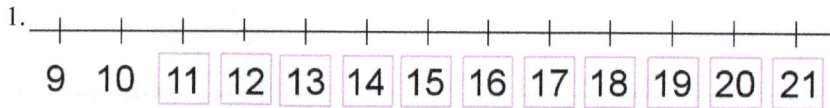

2. a. eleven 11   b. seventeen 17   c. twelve 12   d. sixteen 16
   e. thirteen 13   f. eighteen 18   g. fourteen 14   h. fifteen 15

Page 16

3. a. $10 + 8 = 18$ eighteen   e. $10 + 3 = 13$ thirteen
   b. $10 + 1 = 11$ eleven   f. $10 + 2 = 12$ twelve
   c. $10 + 4 = 14$ fourteen   g. $10 + 7 = 17$ seventeen
   d. $10 + 9 = 19$ nineteen   h. $10 + 5 = 15$ fifteen

4.

1	2	3	4	5	6	7	8	9	10
11	12	13	14	15	16	17	18	19	20
21	22	23	24	25	26	27	28	29	30
31	32	33	34	35	36	37	38	39	40
41	42	43	44	45	46	47	48	49	50
51	52	53	54	55	56	57	58	59	60
61	62	63	64	65	66	67	68	69	70
71	72	73	74	75	76	77	78	79	80
81	82	83	84	85	86	87	88	89	90
91	92	93	94	95	96	97	98	99	100

Building Numbers 11-40, pp. 18-19

Page 18 & 19

1.

a. Eleven	b. Twelve	g. Twenty-one	h. Twenty-two	m. Thirty	n. Thirty-two
1 ten 1 one	1 ten 2 ones	2 tens 1 one	2 tens 2 ones	3 tens 0 ones	3 tens 2 ones
$10 + 1$	$10 + 2$	$20 + 1$	$20 + 2$	$30 + 0$	$30 + 2$
tens ones	tens ones	tens ones	tens ones	tens ones	tens ones
1 1	1 2	2 1	2 2	3 0	3 2
c. Fourteen	d. Fifteen	i. Twenty-three	j. Twenty-six	o. Thirty-five	p. Thirty-seven
1 ten 4 ones	1 ten 5 ones	2 tens 3 ones	2 tens 6 ones	3 tens 5 ones	3 tens 7 ones
$10 + 4$	$10 + 5$	$20 + 3$	$20 + 6$	$30 + 5$	$30 + 7$
tens ones	tens ones	tens ones	tens ones	tens ones	tens ones
1 4	1 5	2 3	2 6	3 5	3 7
e. Seventeen	f. Twenty	k. Twenty-seven	l. Twenty-nine	q. Thirty-eight	r. Forty
1 ten 7 ones	2 tens	2 tens 7 ones	2 tens 9 ones	3 tens 8 ones	4 tens 0 ones
$10 + 7$	$20 + 0$	$20 + 7$	$20 + 9$	$30 + 8$	$40 + 0$
tens ones	tens ones	tens ones	tens ones	tens ones	tens ones
1 7	2 0	2 7	2 9	3 8	4 0

Page 20

1.

a. Forty-one 40 + 1	b. Forty-five 40 + 5	c. Fifty 50 + 0	d. Fifty-seven 50 + 7
tens ones 4 1	tens ones 4 5	tens ones 5 0	tens ones 5 7
e. Seventy-six 70 + 6	f. Eighty-three 80 + 3	g. Ninety-five 90 + 5	h. One hundred 100 + 0 + 0
tens ones 7 6	tens ones 8 3	tens ones 9 5	hundreds tens ones 1 0 0

Page 21

2.

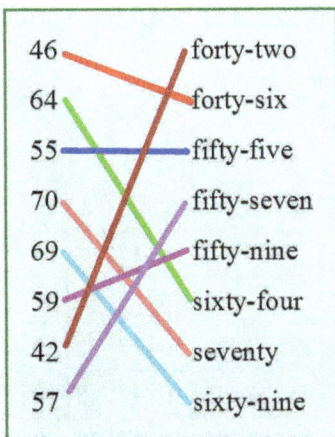 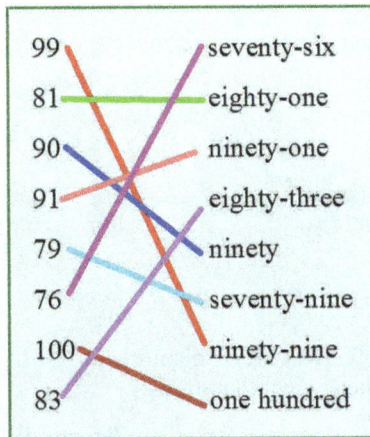

3.

a. 73 = 70 + 3 b. 91 = 90 + 1	c. 45 = 40 + 5 d. 83 = 80 + 3	e. 98 = 90 + 8 f. 64 = 60 + 4

4.

a. 60 + 7 = 67 80 + 0 = 80	b. 4 + 50 = 54 8 + 80 = 88	c. 6 + 80 = 86 0 + 40 = 40	d. 90 + 9 = 99 1 + 60 = 61

5.

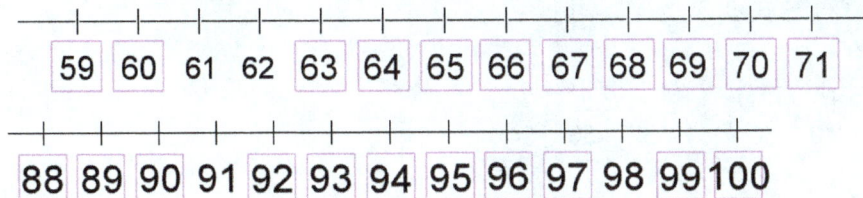

59   60   61   62   63   64   65   66   67   68   69   70   71

88   89   90   91   92   93   94   95   96   97   98   99   100

1 - 2.

1	2	3	4	5	6	7	8	9	10
11	12	13	14	15	16	17	18	19	20
21	22	23	24	25	**26**	27	28	29	30
31	32	33	**34**	35	36	37	38	39	40
41	42	43	44	**45**	46	47	48	49	50
51	52	53	54	55	56	57	**58**	59	60
61	62	63	64	65	66	67	68	69	**70**
71	72	73	74	75	76	77	78	79	80
81	82	83	84	85	86	87	88	89	90
91	92	93	94	95	96	97	98	99	100

4. a. 99, 28   b. 81, 40   c. 85, 69

5. The first nine numbers in the column begin with 3, which represents 3 tens -- the thirty-some things.

6. All of the numbers in that column end in 5, which represents 5 ones.

7. The colored numbers are 28, 53, 76, 82, 100. They are also colored in the 100-chart on the right. Point out to the student that these numbers are right below the given numbers.

8. The underlined numbers are 3, 17, 30, 47, 74, and 89. They are also underlined in the 100-chart on the right. Notice these are directly above the given numbers.

7. and 8. chart

1	2	**3**	4	5	6	7	8	9	10
11	12	**13**	14	15	16	**17**	**18**	19	20
21	22	23	24	25	26	**27**	**28**	29	**30**
31	32	33	34	35	36	37	38	39	**40**
41	42	**43**	44	45	46	**47**	48	49	50
51	52	**53**	54	55	56	**57**	58	59	60
61	62	63	64	65	**66**	67	68	69	70
71	**72**	73	**74**	75	**76**	77	78	79	80
81	**82**	83	**84**	85	86	87	88	**89**	**90**
91	92	93	94	95	96	97	98	**99**	**100**

9. a. 28, 45   b. 70, 69   c. 14, 27

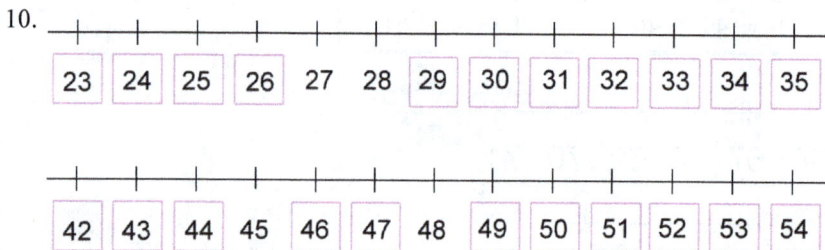

10.

23	24	25	26	27	28	29	30	31	32	33	34	35

42	43	44	45	46	47	48	49	50	51	52	53	54

Add and Subtract Whole Tens, pp. 24-25

Page 24

1. a. 50   b. 80   c. 80   d. 90   e. 100   f. 60
   g. 20   h. 20   i. 30   j. 10   k. 50   l. 60

Page 25

2. a. 60, 80   b. 90, 70   c. 60, 50

3. a. 50, 80   b. 70, 30   c. 10, 60

4. a. 30, 80   b. 50, 40   c. 20, 40

5. a. 50   b. 100

6.

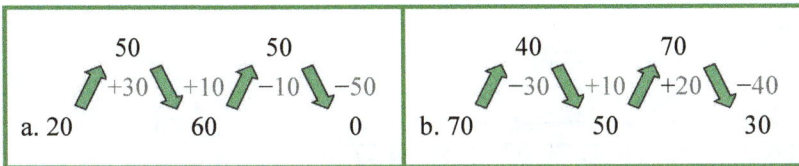

Puzzle corner:

99, 101

Practicing with Numbers, pp. 26-27

Page 26

1.

a. Eighteen
10 + 8
tens | ones
1 | 8

b. Forty-five
40 + 5
tens | ones
4 | 5

c. Sixty-two
60 + 2
tens | ones
6 | 2

d. Thirty-four
30 + 4
tens | ones
3 | 4

2.

a. 43 + 10 = 53
b. 28 + 10 = 38
c. 56 − 10 = 46

3.

| a. 61 + 10 = 71 | b. 37 + 10 = 47 | c. 89 + 10 = 99 |
| 61 − 10 = 51 | 37 − 10 = 27 | 89 − 10 = 79 |

Add and Subtract Whole Tens, cont.

Page 27

4.

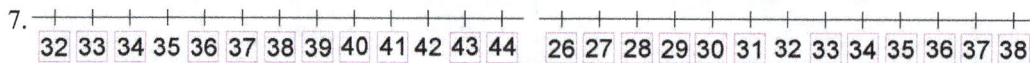

5. a. $91 = 90 + 1$ b. $79 = 70 + 9$ c. $58 = 50 + 8$

6. a. 26, 82   b. 67, 11   c. 56, 35

7.

| 32 | 33 | 34 | 35 | 36 | 37 | 38 | 39 | 40 | 41 | 42 | 43 | 44 |

| 26 | 27 | 28 | 29 | 30 | 31 | 32 | 33 | 34 | 35 | 36 | 37 | 38 |

Mystery Number: 90

Which Number Is Greater?, pp. 28-30

Page 28

1.

a. 63 is more 3 tens 6 ones 6 tens 3 ones 36 < 63	b. 54 is more 5 tens 4 ones 4 tens 5 ones 54 > 45
c. 76 is more 6 tens 7 ones 7 tens 6 ones 67 < 76	d. 64 is more 4 tens 6 ones 6 tens 4 ones 46 < 64

Check first how many __tens__ the numbers have.
If the numbers have the same amount of __tens__, then compare the __ones__.

Page 29

2.

a. 45 < 54	b. 34 > 24	c. 50 < 54
d. 15 < 56	e. 29 < 64	f. 81 < 90
g. 77 > 47	h. 34 < 94	i. 80 > 68

3. a. 80 < 88   b. 95 > 59   c. 99 > 96   d. 85 < 91

e. 90 > 88   f. 49 < 94   g. 90 < 99   h. 100 > 87

i. 50 < 55   j. 22 > 12   k. 98 > 89   l. 24 < 42

58

Which Number Is Greater?, cont.

4. a. 80 + 2 $<$ 80 + 7   b. 10 + 7 $<$ 70 + 5

  c. 60 + 5 $>$ 40 + 5   d. 50 + 7 $<$ 70 + 3

  e. 6 + 60 $<$ 6 + 80   f. 5 + 40 $=$ 40 + 5

  g. 40 + 5 $>$ 40 + 4   h. 70 + 5 $=$ 5 + 70

5. a. 46 < 67 < 68 b. 33 < 37 < 53 c. 18 < 46 < 48 d. 78 < 80 < 87 e. 46 < 48 < 50 < 84 f. 67 < 76 < 87 < 98

Mystery number: 67

Even and Odd Numbers, pp. 31-32

Page 31

1. a. even   b. odd   c. odd
   d. odd   e. even   f. even
   g. odd   h. even   i. odd

Page 32

2.

Cookies	Can you share evenly?	Even or odd?
11	NO	O
14	YES	E
15	NO	O

Cookies	Can you share evenly?	Even or odd?
12	Yes	E
17	No	O
16	Yes	E

Page 32

3.

1	2	3	4	5	6	7	8	9	10
11	12	13	14	15	16	17	18	19	20

21	22	23	24	25	26	27	28	29	30
31	32	33	34	35	36	37	38	39	40
41	42	43	44	45	46	47	48	49	50
51	52	53	54	55	56	57	58	59	60
61	62	63	64	65	66	67	68	69	70
71	72	73	74	75	76	77	78	79	80
81	82	83	84	85	86	87	88	89	90
91	92	93	94	95	96	97	98	99	100

4. Even numbers always end in (their last digit is)
   2, 4, 6, 8, or 0.
   Odd numbers always end in (their last digit is)
   1, 3, 5, 7, or 9.

Numbers Past 100, pp. 33-34

Page 33

1.

101	102	103	104	105	106	107	108	109	110
111	112	113	114	115	116	117	118	119	120
121	122	123	124	125	126	127	128	129	130

2.

81	82	83	84	85	86	87	88	89	90
91	92	93	94	95	96	97	98	99	100
101	102	103	104	105	106	107	108	109	110

Numbers Past 100, cont.

3.

3. a. one more than 109 110 one more than 113 114	b. two more than 116 118 two more than 99 101	c. ten more than 100 110 ten more than 102 112
d. one less than 108 107 one less than 120 119	e. two less than 110 108 two less than 117 115	f. ten less than 120 110 ten less than 107 97

Page 34

4.

a. _one hundred three_

hundreds	tens	ones
1	0	3

b. one hundred six

hundreds	tens	ones
1	0	6

c. one hundred ten

hundreds	tens	ones
1	1	0

d. one hundred fourteen

hundreds	tens	ones
1	1	4

e. one hundred seventeen

hundreds	tens	ones
1	1	7

f. one hundred twenty

hundreds	tens	ones
1	2	0

5.

97	98	99	100	101	102	103	104	105

106	107	108	109	110	111	112	113	114

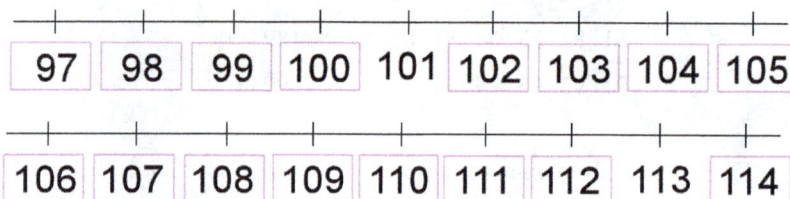

More Practice with Numbers, pp. 35-36

Page 35

1.

31	32	33	34	35	36	37	38	39	40
41	42	43	44	45	46	47	48	49	50
51	52	53	54	55	56	57	58	59	60
61	62	63	64	65	66	67	68	69	70
71	72	73	74	75	76	77	78	79	80
81	82	83	84	85	86	87	88	89	90
91	92	93	94	95	96	97	98	99	100
101	102	103	104	105	106	107	108	109	110
111	112	113	114	115	116	117	118	119	120

More Practice with Numbers, cont.

Page 35

2. a. 77, 97, 55   b. 88, 111, 109   c. 45, 60, 108

Page 36

3. a. 42   b. 54   c. 38

4. a. 61   b. 26   c. 77   d. 84   e. 75   f. 49

5. a. 20 + 20 + 20 = 60
   b. 30 + 30 + 30 = 90
   c. 20 + 20 + 20 + 20 + 20 = 100

6. a. 88, 89, 90, 91   b. 15, 16, 17, 18
   c. 61, 62, 63, 64   d. 107, 108, 109, 110

Skip-Counting Practice, pp. 37-39

Page 37

1. a. 0, 2, 4, 6, 8, 10, 12, 14, 16, 18, 20
   b. 50, 52, 54, 56, 58, 60, 62, 64, 66, 68
   c. 1, 3, 5, 7, 9, 11, 13, 15, 17, 19, 21
   d. 27, 29, 31, 33, 35, 37, 39, 41, 43, 45

2.

1	2	3	4	5	6	7	8	9	10
11	12	13	14	15	16	17	18	19	20
21	22	23	24	25	26	27	28	29	30
31	32	33	34	35	36	37	38	39	40
41	42	43	44	45	46	47	48	49	50
51	52	53	54	55	56	57	58	59	60
61	62	63	64	65	66	67	68	69	70
71	72	73	74	75	76	77	78	79	80
81	82	83	84	85	86	87	88	89	90
91	92	93	94	95	96	97	98	99	100

3. a. 0, 4, 8, 12, 16, 20, 24, 28, 32, 36
   b. 52, 56, 60, 64, 68, 72, 76, 80, 84, 88
   c. 1, 5, 9, 13, 17, 21, 25, 29, 33, 37

Page 38

4.

1	2	3	4	5	6	7	8	9	10
11	12	13	14	15	16	17	18	19	20
21	22	23	24	25	26	27	28	29	30
31	32	33	34	35	36	37	38	39	40
41	42	43	44	45	46	47	48	49	50
51	52	53	54	55	56	57	58	59	60
61	62	63	64	65	66	67	68	69	70
71	72	73	74	75	76	77	78	79	80
81	82	83	84	85	86	87	88	89	90
91	92	93	94	95	96	97	98	99	100

5. a. 0, 10, 20, 30, 40   b. 21, 31, 41, 51, 61
   c. 2, 12, 22, 32, 42   d. 45, 55, 65, 75, 85
   e. 16, 26, 36, 46, 56   f. 59, 69, 79, 89, 99

6. a. 71, 25   b. 44, 92   c. 79, 65

7. a. 0, 5, 10, 15, 20, 25, 30, 35, 40, 45
   b. 1, 6, 11, 16, 21, 26, 31, 36, 41, 46

Mystery Number: 19

Rounding to the Nearest Ten, pp. 40-41

Page 40

1. a. 50   b. 60   c. 40   d. 50   e. 60   f. 50   g. 10   h. 0   i. 10   j. 20   k. 10   l. 0

2. a. 30   b. 50   c. 60   d. 90   e. 10   f. 30   g. 70   h. 90

Page 41

3. a. 40   b. 70   c. 100   d. 80   e. 10   f. 70   g. 80   h. 40

4. a. a skirt ≈ $30, pants ≈ $30, total ≈ $60.
   b. bicycle ≈ $60, light ≈ $10, total ≈ $70.
   c. a puzzle ≈ $20, a book ≈ $10, total ≈ $30

5. About 50 + about 20 is about 70 sacks total

6. About $10 + about $20 + about $20 is about $50.

Bar Graphs, pp. 42-43

Page 42

1. a. five   b. ten   c. 7 + 5 = 12 pencils   d. three more

2. a. 30 students   b. 19 students   c. 24 students

Page 43

3. a. Jane 20   Jerry 35   Jim 15   Hannah 45   Peter 50
   b. Jim read the fewest books.   Peter read the most books.
   c. The two children who read the most books were Hannah and Peter.
      The two children who read the fewest books were Jim and Jane.
   d. Jane and Peter read a total of 70 books. Jim and Hannah read a total of 60 books.

Tally Marks, pp. 44-45

Page 44

1. a. 6   b. 12   c. 19   d. 23

2.

a. 7 卌 ||	b. 14 卌 卌 ||||
c. 16 卌 卌 卌		d. 32 卌 卌 卌 卌 卌 卌 ||
e. 41 卌 卌 卌 卌 卌 卌 卌 卌		f. 28 卌 卌 卌 卌 卌 |||

3.

	Tally Marks	Count
Red	卌		6
Blue	卌 卌	10
Yellow	卌 卌		11

Page 45

4.

	Tally Marks	Count
Group 1	卌 |||	8
Group 2	卌 卌		11

5. The tally marks for blue, white, and red cars were in error.

	Tally Marks
Blue	卌 卌 卌 卌 卌
Black	卌 卌
White	卌 卌 卌 卌 卌 卌
Red	卌 卌 卌 卌

Regrouping: Make Another Ten from the Ones, pp. 46-49

Page 46

Answers will vary in the sense that it does not matter which exact one-dots are circled.

1.

a. Three times 16 = <u>48</u>

b. Four times 13 = <u>52</u>

c. Three times 23 = <u>69</u>

d. Two times 35 = <u>70</u>

e. Two times 28 = <u>56</u>

f. Two times <u>17</u> = <u>34</u>

g. Three times <u>36</u> = <u>108</u>

h. Three times <u>25</u> = <u>75</u>

i. Three times <u>15</u> = <u>45</u>

Page 47

2.

a. Two times 19 = <u>38</u>

b. Two times 34 = <u>68</u>

c. Three times 26 = <u>78</u>

d. Four times 17 = <u>68</u>

e. Three times 29 = <u>87</u>

f. Two times 38 = <u>76</u>

Regrouping: Make Another Ten from the Ones, cont.

Page 48

3.

a. 14 + 15 + 16 = 45	b. 21 + 26 + 17 = 64
c. 28 + 25 = 53	d. 36 + 26 = 62
e. 18 + 16 + 14 = 48	f. 24 + 26 + 18 = 68

Page 49

4.

a. 17 + 26 = 43	b. 23 + 28 = 51
c. 22 + 18 = 40	d. 19 + 12 + 16 = 47

Regrouping: Make Another Ten from the Ones, cont.

Page 49

4. (continued)

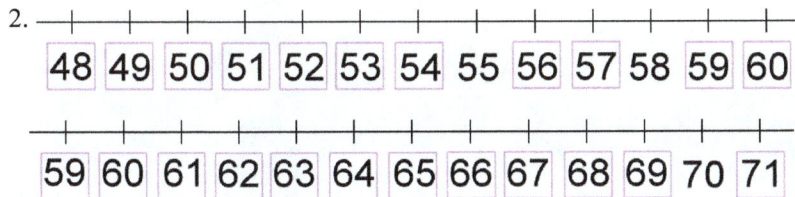

e. 37 + 15 = 52

f. 20 + 13 + 19 = 52

Review, pp. 50-51

Page 50

1. a. 1 ten 5 ones   _15_   fifteen
   b. 6 tens 7 ones   67   sixty-seven
   c. 4 tens 0 ones   40   forty
   d. 10 tens 0 ones   100   one hundred
   e. 5 tens 1 one   51   fifty-one

2.

| 48 | 49 | 50 | 51 | 52 | 53 | 54 | 55 | 56 | 57 | 58 | 59 | 60 |

| 59 | 60 | 61 | 62 | 63 | 64 | 65 | 66 | 67 | 68 | 69 | 70 | 71 |

3. a. 87   b. 25   c. 57   d. 80   e. 101

4. 97, 98, 99, 100, 101, 102, 103, 104, 105, 106, 107, 108, 109

Page 51

5.

a. 45 = 40 + 5	b. 25 = 20 + 5	c. 78 = 70 + 8
68 = 60 + 8	54 = 50 + 4	91 = 90 + 1

6.

a. 50 + 7 = 57	b. 8 + 10 = 18	c. 90 + 6 = 96
20 + 0 = 20	9 + 70 = 79	9 + 60 = 69

7. a. 17 < 57 < 75   b. 18 < 41 < 44 < 48

8. a. 56 $<$ 5 + 60   b. 20 + 8 $<$ 33   c. 60 + 5 $>$ 50 + 6

   d. 34 $<$ 30 + 6   e. 4 + 90 $>$ 49   f. 80 + 2 $>$ 70 + 9

9. a. 13, 15, 17, 19, 21, 23, 25, 27, 29
   b. 18, 28, 38, 48, 58, 68, 78, 88, 98
   c. 30, 35, 40, 45, 50, 55, 60, 65, 70

Mystery Number: 54

On the following page is a extra number chart and empty number lines for you to print as needed.

Number Chart _____

Number Lines

More from math MAMMOTH

Math Mammoth has a variety of resources to fit your needs. All are available as economical downloads, and most also as printed copies.

- **Math Mammoth Light Blue Series**
  A complete curriculum for grades 1-7. Each grade level includes two student worktexts (A and B), which contain all the instruction and exercises all in the same book, answer keys, tests, cumulative reviews, and a worksheet maker. International (all metric), Canadian, and South African versions are also available.

  https://www.MathMammoth.com/complete-curriculum

  https://www.MathMammoth.com/international/international

  https://www.MathMammoth.com/canada/

  https://www.MathMammoth.com/south_africa/

- **Math Mammoth Skills Review Workbooks**
  These workbooks are intended to be used alongside the Light Blue series full curriculum, and they provide additional review to the topics studied in the main curriculum, in a spiral manner.
  https://www.MathMammoth.com/skills_review_workbooks/

- **Math Mammoth Blue Series**
  Blue Series books are topical worktexts for grades 1-7, containing both instruction and exercises. The topics cover all elementary mathematics from 1st through 7th grade. These books are not tied to grade levels, and are thus great for filling in gaps.
  https://www.MathMammoth.com/blue-series

- **Make It Real Learning**
  These activity workbooks concentrate on answering the question, "Where is math used in real life?" The series includes various workbooks for grades 3-12.
  https://www.MathMammoth.com/worksheets/mirl/

- **Review Workbooks**
  Workbooks for grades 1-7 that provide a comprehensive review of one grade level of math—for example, for review during school break or summer vacation.
  https://www.MathMammoth.com/review_workbooks/

Free gift!

- Receive over 350 free sample pages and worksheets from my books, plus other freebies:
  https://www.MathMammoth.com/worksheets/free

Lastly...

- Inspire4 is an inspirational website for the whole family I've been privileged to help with:
  https://www.inspire4.com